INSTRUCTOR'S MANUAL
for
ELECTRIC MOTORS
Principles, Controls, Service and Maintenance

TABLE OF CONTENTS

UNIT	PAGE
INTRODUCTION	1
INSTRUCTOR LEVEL OF COMPETENCY	3
STUDENT LEVEL OF COMPETENCY	4
STUDENT MANUAL TABLE OF CONTENTS	5
PLANNING SCHEDULE FOR INSTRUCTIONAL UNITS	
INSTRUCTION UNITS FOR GROUPS	7
SLIDE/CASSETTE PROGRAMS	8
WORKSHEETS	8
TRANSPARENCIES AND TRANSPARENCY MASTERS	8
HARDWARE FOR INSTRUCTIONAL UNITS	
TOOLS, EQUIPMENT AND KITS	9
SUPPLIES FOR ALL INSTRUCTIONAL PROGRAMS	10
SAFETY EQUIPMENT FOR INSTRUCTIONAL UNITS	10
SUPPLEMENTAL REFERENCES FOR INSTRUCTIONAL UNITS	10
OHMS LAW APPLICATIONS	11
EXPERIMENTAL MOTORS	15
ANSWERS: CLASSROOM AND LABORATORY ACTIVITIES	23

AUTHORS

W. FORREST BEAR
Professor
Departments of Agricultural Engineering
and Vocational and Technical Education
University of Minnesota
St. Paul, MN 55108

HARRY J. HOERNER
Professor
Department of Agriculture
Western Illinois University
Macomb, IL 61455

HOBAR PUBLICATIONS
DIVISION OF HOBAR ENTERPRISES, INC
1234 Tiller Lane
St. Paul, MN 55112

INTRODUCTION

This manual, <u>Electric Motors: Principles, Controls, Service and Maintenance</u> has been written with the intent to satisfy several instructional areas with different lengths of instructional programs.

In the review of available manuals the authors felt there are two basic types: those written by the engineers at a technical level and those written by educators which are less technical and primarily for agricultural and industrial education students. Each electric motor company publishes technical data sheets plus service and maintenance information on their electric motors. The title of this manual indicates a combination of these subject matter sources which is essential for the intended audience. A concern of the authors is the technical level selected. We intend to have raised the lower educational limit and simplified the upper educational level in such a way that our audience will be more intelligent about the understanding and utilization of electric motors.

The number of electric motors in our homes, businesses and factories is staggering in number. As more and more machines are powered by motors the more competent each of should be for the selection, purchase, service and maintenance of the electric motor. The best electric motor can be ruined if any one of the previously stated competencies is not considered with the correct management decision.

The authors have taught electric motors to high school students and adults plus college students and vocational instructors in methods of teaching courses. These teaching experiences have prompted us to write this manual and accompanying visual aids. Possibly, what we have not known as instructors has been an incentive for us to prepare a better instructional package for electric motors.

If we have accomplished this and have helped you plan and teach a better instructional program on electric motors our goal will have been achieved.

One reviewer of this manual could not imagine this many words, figures and tables on electric motors. He did not explain whether he anticipated more or less information. The authors feel this manual can be used to teach more than just electric motors. It can be the basis for data on wiring service, generation of electricity, overload protection, switches, controls plus the electric motor information. The electrical principles covered are involving motors in this manual, but many of these electrical principles could have been directed to other types of electrical loads, such as in heating and lighting.

The authors are expecting that before students study this manual, they will have achieved the subject matter in the manual, <u>Basic Electricity and Practical Wiring</u>. If the <u>Basic Electricity and Practical Wiring</u> text has not been used, the instructor can cover the subject matter in the beginning or incorporate it into this unit of electric motors.

This manual sets the stage for a basic understanding of electrical principles and commonly used wiring practices. For example, an understanding of parallel and series circuit and electrical terms is a must if you are to understand motors.

Each unit will have one or more classroom and/or laboratory exercises for the instructional programs. Many teachers like to have special projects to construct. Hobar Publications has made special arrangements with Delmar Publications to include five experimental electric motors in this publication. Although they are simple electrically, there is a certain amount of expertise and patience required to construct them.

The authors desire is that we have created an interest in electric motors and that you will be challenged to have a better instructional program.

INSTRUCTOR LEVEL OF COMPETENCY

How much should the instructor know to teach a unit on electric motors? Whether you are teaching 5, 10, 15, 30, 60 or 90 hours to high school students, whether you are an instructor at an area vocational technical institute or at the university level, you can't know too much. Regardless of your level of technical competence and practical experience achieved there will still be students in your class who have a competence that you have not achieved and will ask questions which you cannot answer. This situation promotes the purchase of technical references and develops a new friendship with the owner and technicians at the local electric motor rewind and repair shop. Such establishments may also be used as community resources for educational field trips and expert guest speakers to the classes studying electric motors. Part of instruction in any vocational and/or technical area is to cause students to become occupationally aware of industries and future jobs. In fact, the authors believe a well taught electrical motors unit may spark an occupational awareness in a number of the students who may desire to work this exciting and good paying occupational field.

Do not wait until you know everything about electric motors before you teach a unit. If so, you will never make this addition to your curriculum. Electric motors are here to stay and they represent a power unit that is very influential in our daily lives. Electric motors are relatively small and readily available without a great expense. The unit can be taught with or without an extensive laboratory facility. This instructional unit will provide students with a knowledge and competence that will be applicable throughout life and open a bright future for those excited about this subject matter.

STUDENT LEVEL OF COMPETENCY

How much should the student know about electric motors? When electric motors are working they are usually out-of-sight and out-of-mind. When they are not working you are deprived of the use of a power tool, the furnace, the water from a well, the food mixer, the washing machine, the sump pump or similar motor powered applications in your home. Every business and factory utilizes motors. Therefore, your quality of life and livelihood may depend on electric motors.

The authors don't know exactly what students should know about electric motors. We are confident that all the subject matter in the manual, <u>Electric Motors: Principles, Controls, Service and Maintenance</u> will be advantageous to them. Some students will have a need for selection and installation information; others will need service and repair assistance; and, others will need the technical information about the operation of electric motors.

Regardless of the student's goal, it is the authors desire that our manual, classroom and laboratory exercises will spark an interest which will make him/her want to learn more about electric motors. As a possible career, electric motors and related work, has an outstanding future for those persons with the interest and training in this area of study.

The table of contents for the student manual follows and displays the extent of the subject matter covered in the manual.

TABLE OF CONTENTS

ELECTRIC MOTORS
Principles, Controls, Service and Maintenance

Unit	Page
I. HISTORY AND APPLICATION OF ELECTRIC MOTORS	5
Development of Electric Motors	5
Magnetism	5
Electricity	5
Induction	5
Electromagnet	5
Generation of Electricity	5
DC Motors	5
Alternating Current	6
AC Motors	6
Electric Motors Today	6
Definition of Terms	8
Classroom Exercises	9
II. EXTERNAL FEATURES OF ELECTRIC MOTORS	11
Mount and Base	11
End Bells and Enclosures	13
Bearings, Shafts and Lubrication	14
Cooling Systems	17
Mounting Position	18
Definition of Terms	19
Classroom Exercises	21
Laboratory Exercises	24
III. NAMEPLATE INFORMATION	25
Electrical Features	25
Horsepower	25
Type of Current	26
Phase	26
Cycles or Hertz	27
Revolutions Per Minute	27
Volts	28
Amperes	28
Thermal Protection	28
Code	30
Service Factor	30
Time	31
Efficiency Index	31
Physical Features	31
Insulation Class	31
Temperature Rise	32
NEMA Design	33
Frame	34
Enclosures and End Bell Ventilation	36
Shaft Diameter	38
Bearing	38
Manufacturer's Designations	38
Type	38
Identification Number	38
Enclosures	39
Model Numbers	42

Unit	Page
Definition of Terms	43
Classroom Exercises	44
Laboratory Exercises	50
IV. PRODUCTION OF ELECTRICAL ENERGY	53
Direct Current	53
Alternating Current, Single-Phase	53
Alternating Current, Three-Phase	54
Voltage Supplied	54
Delta Transformer	55
Wye "Y" or Star Transformer	56
Definition of Terms	57
Classroom Exercises	58
V. MOTOR CLASSIFICATION AND OPERATION	61
Motor Classification	61
Motor Parts and Circuits	61
Magnetism and Electricity	61
Motor Running Circuit	64
Definition of Terms	66
Classroom Exercises	68
Laboratory Exercises	70
VI. STARTING SYSTEMS AND CIRCUITS	71
Alternating Current Induction-Run Motors	71
Split-Phase-Start Motors	71
Capacitor-Start Motors	73
Shaded Pole-Start Motors	73
Repulsion-Start Motors	74
Alternating Current Single-Phase Synchronous Motors	75
Alternating Current Universal Motors	75
Alternating Current Three-Phase Motors	76
Definition of Terms	78
Classroom Exercises	80
Laboratory Exercises	82
VII. CHANGING VOLTAGE, REVERSING ROTATION AND CHANGING MOTOR SPEED	83
Changing Voltage Service	83
Reversing Motor Rotation	84
Changing Motor Speed	85
Split Phase Motor, Two Speed	86
Permanent-Split Capacitor	86
Definition of Terms	88
Classroom Exercises	89
Laboratory Exercises	90
VIII. THE MOTORS WE USE	91
Induction-Run Type Motors	91
Wound Rotor Motors	94
Definition of Terms	96
Classroom Exercises	97
Laboratory Exercises	98

ELECTRIC MOTORS
Principles, Controls, Service and Maintenance

Unit	Page

IX. ELECTRIC MOTOR SELECTION AND PERFORMANCE TESTING 99

    Types of Loads for Motors 99

        Continuous Running Steady Loads 99
        Continuous Running Intermittent Loads 99
        Adjustable Speed Loads 99
        Cyclical Loads 100
        Motor Characteristics 100

    Performance Testing 103

        Equipment and Test Meters 103
        Safety Rules for Data Collection 104
        Formulas and Meter Readings..............104
        Analysis of Data 106

    Motor Performance and Design................107

        Speed Represented as RPM.................109

    Definition of Terms 110
    Classroom Exercises 111
    Laboratory Exercises 113

X. ELECTRICAL SERVICE AND CONTORL DEVICES 119

    Electric Motor Circuit 119

        Size of Conductor 119
        Running Overcurrent Protection 119
        Circuit Overcurrent Protection 121
        Overcurrent Devices 121

    Nonautomatic Devices for Controlling Motors 122

        Switches without Overload Protection........122
        Switches with Overload Protection 124

    Motor Controllers...........................126
    Automatic Control Devices for Controlling Motors 131
    Definition of Terms 134
    Classroom Exercises 136
    Laboratory Exercises 142

XI. MOTOR INSTALLATION 143

    Environmental Factors 143

        Dust 143
        Water or Moisture 143
        Stray Oil 143
        Air....................................143

    Alignment 144
    Vibration 145
    Drive System 145

        Chain and Gear Drives 147
        Direct Drive 149

    Definition of Terms 150
    Classroom Exercises 151
    Laboratory Exercises 154

XII. ELECTRIC MOTOR MAINTENANCE..........155

    Analyzing the Situation 155
    Bearing Maintenance........................156
    Cleaning the Motor 157
    Electric Motor Service and Repair 159
    Definition of Terms 168
    Laboratory Exercises 170

XIII TROUBLE SHOOTING 173

    Classroom Exercises 178
    Laboratory Exercises 179

APPENDIX 180

    References..................................180

    Constructing Experimental Motors 183

    Data Tables.................................189

INDEX 197

SUGGESTED INSTRUCTIONAL OUTLINE

INSTRUCTIONAL GROUPS	UNITS FOR						POST SECONDARY AVTI	UNIVERSITY	
UNIT NUMBER	5 HRS	10 HRS	15 HRS	30 HRS	60 HRS	90 HRS		OPTION I	OPTION II
1. History and Application	GENERAL INTEREST FOR ALL GROUPS								
2. External Features									
3. Nameplate Information									
4. Production of Electrical Energy									
5. Motor Classification and Operation									
6. Starting System and Circuit									
7. Change Voltage, Reverse, Change Speed									
8. The motors We Use									
9. Selection & Performance Test									
10. Electrical Service & Control Devices									
11. Motor Installation									
12. Electric Motor Maintenance									
13. Trouble Shooting									

PLAN YOUR ELECTRIC MOTOR INSTRUCTIONAL
UNIT ON THIS PAGE

SLIDE/CASSETTE PROGRAMS FOR INSTRUCTIONAL UNITS

Title and Hobar Publications order number	Recommended Unit and/or Group
Electric Motors: Enclosures, Bearing and Mounting Methods, #471	2
Electric Motors, Service and Maintenance, #452	12
Electrical and Basic Controls, #457	10
Alternating Current Electric Motors, #514	all groups
Electric Motor Nameplate Data, #515	3
Electric Motor Speed Determination, and Torque Curve Characteristics, #516	5, 6, 7
Electric Motor Application and Characteristics, #517	5, 8
How Electric Motors Start and Run, #AV300F (filmstrips)	5, 6

WORKSHEETS FOR INSTRUCTIONAL UNITS

Electric and Basic Controls Worksheets, #863	10
Electric Motor Worksheet Set, #758, for the set of 20	all groups

TRANSPARENCIES AND TRANSPARENCY MASTERS FOR INSTRUCTIONAL UNITS

Electric and Basic Controls, #2277	10
Electric Motor Transparencies, #183TM	all units

TOOLS AND EQUIPMENT FOR ALL INSTRUCTIONAL PROGRAMS

Basic Tools and Kits	0-15 Hrs.	15-30 Hrs.	30+ Hrs.
Electric Motor Tool Kit, EMK-1	x	x	x
Electrical Tool Kit, #ETK-76	x	x	x
Clamp-On Test Meter, #G4X221	x	x	x
Kilowatt Hour Meter, #KWHM-1			x
Solderless terminal kit, #G4X258	x	x	x
Wire stripper, #18-800	x	x	x
Soldering gun, #G2Z777	x	x	x
Nut driver set, #IND-1			x
¼" Socket set, #SK4913	x	x	x
3/8" Socket set, #SK4512	x	x	x
Gear Puller - Z - 630	x	x	x
Bearing Separator	x	x	
Basic Electric and Control Kit, #EBC-1			x
Watt Meter			x
Line reamer set			x
Metal lathe			x
Hydraulic press			x
Continuity tester, #BE61	x	x	x
Neon circuit tester, #ET-1	x	x	x
Ignition point file, #Cal Van 608	x	x	x
Oil can, #6X810	x	x	x
Grease gun, #6X808	x	x	x
Wood dowel, Local Building Supplier	x	x	x
GFCI extension cord, #G6X150	x	x	x
Outside Micrometer, 0-1", ST 02 0851 04	x	x	x
Hanging spring scales for testing			x
Vernier Caliper, ST 52 1341 00		x	x
Dial indicator, #ST 50 0133 00		x	x
Magnetic base for dial indicator, ST 59 0001 00		x	x
Capacitor start motor, #6K081		x	x
Split phase-start motor, #5K533		x	x
Tachometer, 0-4000rpm, #TM-1	x	x	x
Vibration tachometer, #TM-2	x	x	x
Impact Wrench, #IT-1		x	x
Hex wrench pak, #HPI-1	x	x	x
Pneumatic tools, #R606 or R644		x	x
Heater cones (600 w) and bases		x	x
Diagonal cutter DC-6	x	x	x

SUPPLIES FOR ALL INSTRUCTIONAL PROGRAMS

1. Shims and thrust washers
2. Fine and medium grit emery cloth
3. Emery boards
4. Miscellaneous sleeving
5. Fiberglass tape

6. Fiberglass ties
7. Recommended sleeve bearing lubricating oil
8. Recommended grease
9. Miscellaneous #6, 8, and 10 National Fine and National Coarse Nuts, # 620R
10. USS flat washers, #394J

11. Resin core tin lead soft solder 40/60, #2Z173
12. Silver solder and correct flux
13. Electric contact spray, #4X598
14. Spray paint, glossy black, #2X711
15. Lead wire, stranded, 14, 16 or 18 as recommended

16. Wiring terminals and connectors, #10-003, 10-004 and 10-006
17. Power cord, SJO 14/3
18. Dead front plugs, #H5266N
19. Penetrating oil, #WD40
20. SAE flat washers assortment, #618J

21. Spring lock washers assortment, #745J
22. Hex brass machine screw nuts assortment, #621R
23. Round head slotted stove bolt assortment, #270J
24. Round head slotted machine screw assortment, #895J
25. Spray belt dressing, #2X987

SAFETY EQUIPMENT

1. First aid kit, #5X624
2. Safety eyewear, Style B, C720W or C720S
3. Flexible cover goggles, # 880 LFP

SUPPLEMENTAL REFERENCES

1. Basic Electricity and Practical Wiring, #2377 and instructor book, #3177
2. Electrical and Basic Controls used in Agricultural Production, #2177
3. National Electrical Code Book #380
4. Agricultural Wiring Handbook, #574
5. Introduction to Electricity and Electronics, #1160-5 and instructor book #1162-1

OHMS LAW APPLICATIONS

I. The intensity of the current in amperes in an electric circuit is equal to the difference in potential in volts across the circuit divided by the resistance in ohms of the circuit.

   A. I = V/R

       1. This is exact when applied to D.C. circuits.
       2. It is true in a modified form for A.C. circuits.

   B. Current varies inversely with the resistance when voltage is constant.

       I = V/R; = 12/2 = 6: 12/3 = 4: 12/4 = 3: 12/6 = 2.

       1. Thus, current decreases as resistance increases, voltage constant.

   C. Current varies directly with the voltage when the resistance is constant.

       I = V/R; = 12/3 = 4: 15/3 = 5: 18/3 = 6: 21/3 = 7.

       1. Thus, current increases as voltage increases, resistance constant.

   D. Voltage varies directly with the resistance when amperage is constant.

       V = I x R; = 3 x 4 = 12: 3 x 5 = 15: 3 x 6 = 18: 3 x 7 = 21.

       Thus, voltage increases as resistance increases, amperage constant.

   E.

SERIES CIRCUITS

I. If a circuit is arranged so the current has only one possible path from the voltage source through conductors to the load or loads and back to the voltage source, it is a series circuit.

    A. To find the current in a series circuit.

        1. Find the total resistance.

            a. $R_t = R_1 + R_2 + R_3 + R_4$ etc.

        2. Then use: $I = \dfrac{V}{R_t}$

    B. To find the voltage required in a series circuit when the various resistance values and current are known.

        1. Find the total resistances.

        2. Then use: $V = I \times R$

    C. To find the voltage drop through a resistance when the current and the resistance are known.

        1. $V = I \times R$

    D. To find the total voltage when the voltage drops through the circuit are known.

        1. $V_t = V_1 + V_2 + V_3 + V_4 + V_5$ etc.

II. Current in a series circuit is the same in all parts of the circuit. For a given voltage, the current is determined by the total resistance of the circuit.

$$I_t = I_1 = I_2 = I_3 \ldots$$

III. Total resistance of a series circuit is equal to the sum of the various resistances in the circuit.

$$R_t = R_1 + R_2 + R_3 \ldots$$

IV. Total voltage drop in a series circuit is equal to the sum of the voltage drops through the various resistances. Voltage drop through each resistance is equal to the current times the resistance.

$$V_t = V_1 + V_2 + V_3 \ldots$$

PARALLEL CIRCUITS

I. If a circuit is arranged so the current has more than one possible path from the voltage source through conductors to the load or loads and back to the voltage source, it is a parallel circuit.

   A. To find the resistance of a parallel circuit when the values of the individual resistances are known.

   1. $R_j = \dfrac{R_1 \times R_2}{R_1 + R_2}$ (for 2 resistances) and

   2. $\dfrac{1}{R_t} = \dfrac{1}{R_1} + \dfrac{1}{R_2} + \dfrac{1}{R_3}$ (for more than 2) R total is always less than the smallest R values.

   B. To find the current in a parallel circuit.
      1. Find the joint resistance.
      2. Then use: $I = \dfrac{V}{R}$

   C. To find the voltage required to force a certain current through a parallel circuit containing two resistances.
      1. Find the joint resistance
      2. Then use: $V = I \times R$

   D. To find the current flowing through one resistance in a parallel circuit when the current through the other resistance is known.
      1. Find voltage: (voltage is the same at each resistance in a parallel circuit)
      2. Find current: $I = \dfrac{V}{R}$

II. Current in each branch of a parallel circuit is equal to the voltage divided by the resistance of the branch. The total current in a parallel circuit is equal to the sum of the currents in the individual branches.

$$I_t = I_1 + I_2 + I_3 \ldots$$

III. The voltage in a parallel circuit is the same at all branches of the circuit or the voltage at each branch of a parallel combination is the same as the voltage in the circuit.

$$V_t = V_1 = V_2 = V_3 \ldots$$

IV. Resistance of a parallel circuit is equal to the voltage divided by the sum of the currents in the individual branches. See item IA.

SERIES-PARALLEL CIRCUITS

I. To find the voltage at a parallel combination when there is a resistance in series.

   A. First, the voltage drop through the series resistance must be known.

      1. To find this drop value of the resistance.

      2. Subtract the voltage from the total voltage will give the voltage at the parallel combination.

II. To find the total resistance in a series-parallel circuit.

   A. Total resistance is equal to the sum of the resistance in the series and the joint resistance of the parallel group.

III. To find the current through each parallel resistance.

   A. If resistance is given and the voltage of the circuit is known: I = V ÷ R

USE THIS PAGE FOR NOTES ON OHMS LAW APPLICATIONS
AND SERIES & PARALLEL CIRCUIT EXAMPLE PROBLEMS

EXPERIMENTAL MOTOR NO. 2

This motor, figure 13-36, with a three-pole armature, is self-starting and has considerable torque for its size. It may be connected to operate as a series motor or as a shunt motor on 6-volt direct current.

MATERIALS

48 - Pcs. #22 gauge sheet iron
 1 - Pc. CRS 1/8" x 3 1/2" (3 mm x 89 mm)
 1 - Bolt, 1/2" x 3" (M12 x 75 mm L)
 1 - Nut, 1/2" (M12)
 1 - Machine screw, #10-32 x 1" (M5 x 24 mmL)
 2 - Pcs. sheet iron, #22 gauge, 1" x 2 3/4" (25 mm x 70 mm)
 1 - Pc. Copper-clad phenolic board, 1 1/4" (32 mm) diameter
 2 - Pcs. Spring brass wire, #16 gauge x 4" (100 mm)
 4 - Pcs. soft steel, 1/8" x 1" x 4 1/4" (3 mm x 25 mm x 108 mm)
 2 - Fiber washers, 1/16" x 1 1/2" diameter (1.5 mm x 38 mm diameter)
 1 - Wooden base, 3/4" x 3 3/4" x 6" (20 mm x 95 mm x 150 mm)
 2 - Fahnestock clips
 1 - Washer, #6 (M3.5)
 4 - Wood screws, #4 x 1/2", RH
 3 - Wood screws, #6 x 1/2", RH
 1 - Pc. tubing, 1/8" ID x 1/4" (3 mm ID x 6 mm)
 1 - Pc. tubing, 1/8" ID x 3/8" (3 mm ID x 9 mm)
 2 - Nuts, #5-44 (M3)
 1 - Fiber washer, 1/2" (12 mm) diameter with 1/8" (3 mm) hole
 1 - Pulley
    Magnet wire, #22 gauge AWG Formvar

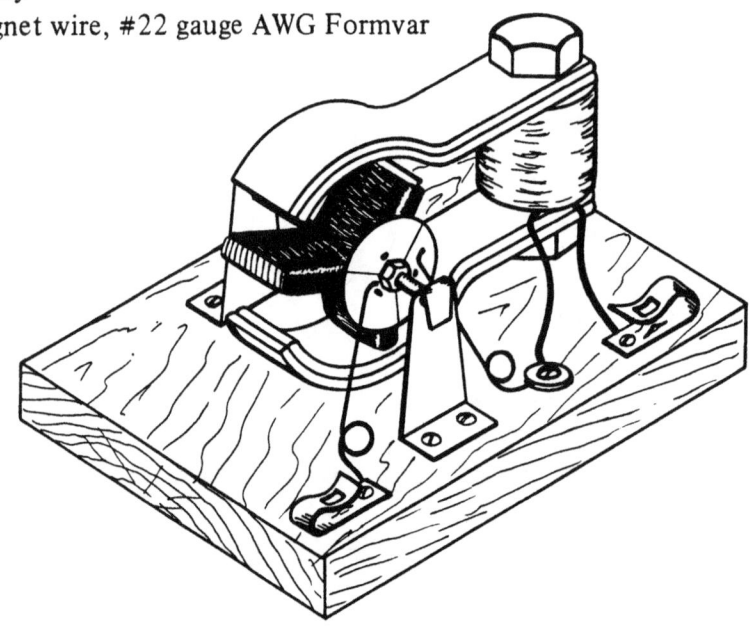

Fig. 13-36

PROCEDURE

Study the drawings, figures 13-37 and 13-38, and the suggested procedure. Procure the materials and make the parts.

When cutting the armature laminations, be careful to keep the metal flat.

Remove all burrs, stack the pieces evenly, clamp them together, and drill the hole through the center. Select a drill which will produce a tight fit between the laminations and the shaft.

Insert the shaft and balance the rotor.

Before winding the armature coils, round the corners of each rotor pole with a fine file. Next, wrap two layers of plastic tape around each one. Starting at the center each time, wind four layers of #22 magnet wire on each pole. Wind each coil in the same direction and with the same number of turns.

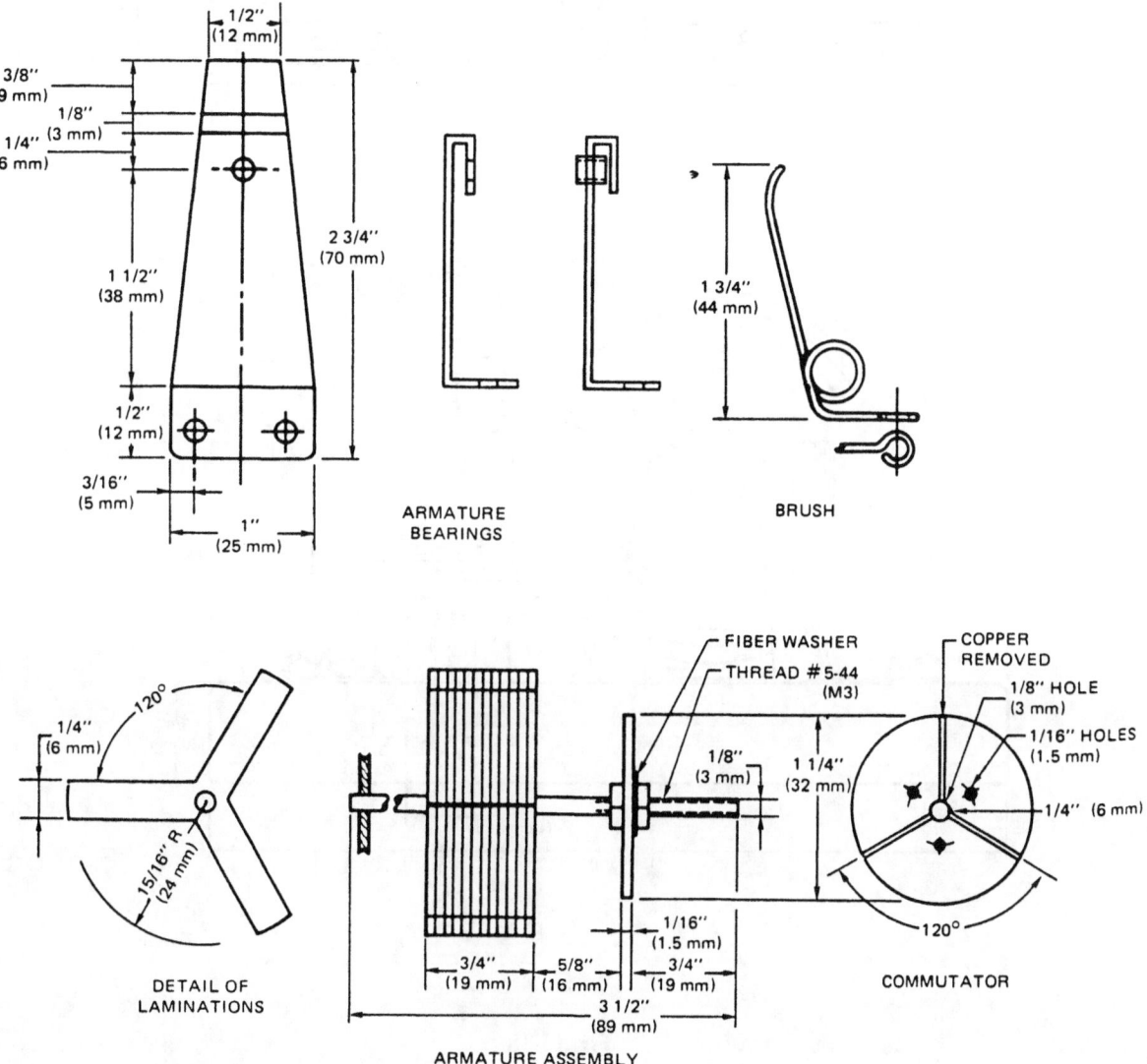

Fig. 13-37

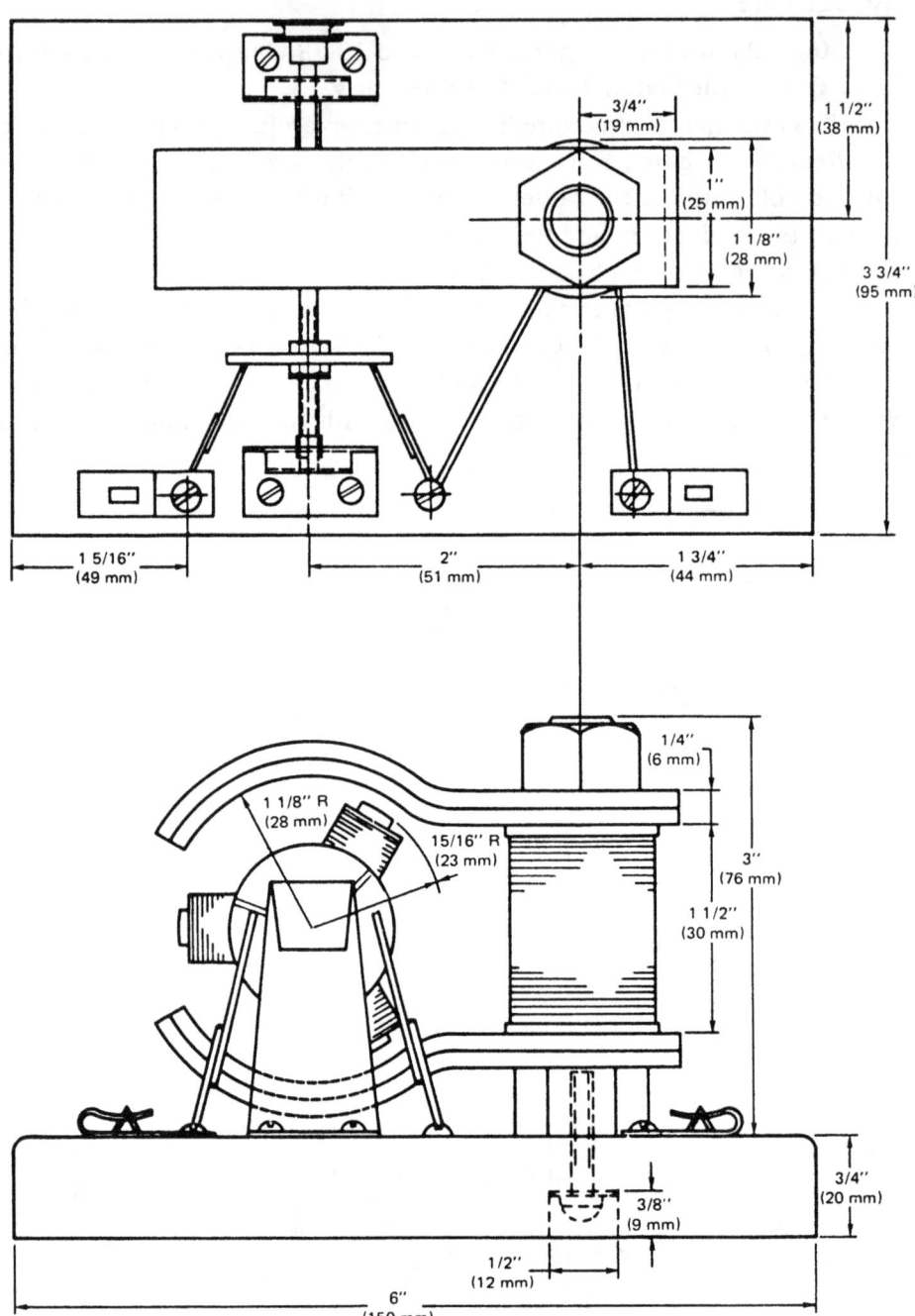

Fig. 13-38

Leave sufficient wire on each end to reach the commutator. Secure the ends with cotton tape.

Check each coil for ground and coat with shellac or varnish.

The field coil is wound on the bolt over insulating material. It consists of 8 layers of #22 magnet wire. When winding the coil, start from the nut end. Leave enough wire on each end to make connections.

The bolt is drilled and tapped so the frame and coil can be fastened to the base with a machine screw.

The field pieces may be made from either 1/16" (1.5 mm) or 1/8" (3 mm) x 1" (25 mm) band iron or soft steel. If 1/16" (1.5 mm) material is used, eight pieces will be required.

Since the shaft is threaded on the commutator side, the armature bearing is fitted with a short piece of tubing to prevent the threads from binding.

The commutator is made of copper-clad phenolic board and secured to the shaft with two nuts. A fiber washer is placed between the disc and the nut on the copper-clad side to prevent a short. Enough copper should be removed from each segment at the edge of the hole to prevent it from touching the shaft.

The armature coils are connected so the beginning end of each is fastened to a commutator segment along with the finish end of the coil to the right of it. The wires are threaded through holes in the commutator and soldered to the copper face.

Note: A three-segment commutator may be made with tubing.

The brushes are formed from brass spring wire. They are mounted to the base with wood screws. Care should be taken to adjust them for the proper tension.

Assemble the parts and test. The position of the commutator segments in relation to the poles may need to be adjusted to provide maximum torque.

Evaluate.

All information for the experimental electric motors:
From Loper, Orla; Ahr, Arthur; and Glendenning, Lee. Introduction to Electricity and Electronics (c) 1979 by Delmar Publications, Inc. Used with permission of Delmar Publishers, Inc.

DESIGN AND BUILD A SYNCHRONOUS MOTOR

This simple synchronous motor, figure 16-30, is comparatively easy to build.

It is designed to run on alternating current from a low-voltage transformer.

It is not self-starting and has very little torque.

Its principal use is to illustrate the constant speed characteristic of a synchronous motor.

The rpm of the motor is determined by the number of poles on the rotor.

MATERIALS

1 - Machine screw, #8-32 x 2" (M4 x 50 mm L)
2 - Machine screw nuts, #8-32 (M4)
1 - Machine bolt, 3/8" x 2 1/2" (M10 x 65 mm L)
1 - Machine bolt nut, 3/8" (M10)
2 - Fiber discs, 1/16" x 1" (1.5 mm x 25 mm)
2 - Pieces soft steel, 1/8" x 1/2" x 2 1/2" (3 mm x 12 mm x 64 mm)
1 - Piece soft steel, 1/8" x 1" x 2 1/4" (3 mm x 25 mm x 57 mm)
2 - Fahnestock clips
3 - Pieces sheet iron, #18, #20, or #22 gauge, 3/8" x 3" (9 mm x 76 mm), or 1 piece 3" x 3" (76 mm x 76 mm)
1 - Piece wood, 3/4" x 2" x 5 1/4" (20 mm x 50 mm x 133 mm)
2 - Wood screws, #6 x 1" RH
2 - Wood screws, #5 x 3/4" RH
2 - Wood screws, #4 x 1/2" RH
   Magnet wire, #24 gauge, AWG Formvar

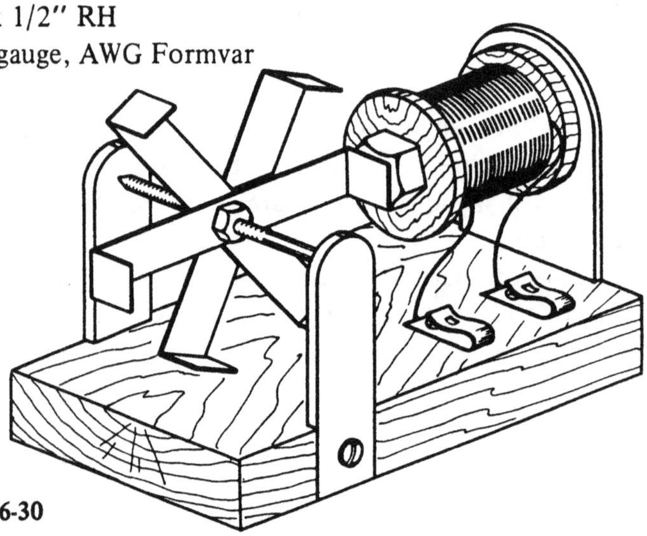

Fig. 16-30

PROCEDURE

Study the relationship of the various parts in figure 16-31 and make each according to the dimensions given in the materials list.

Plan the coil to be 2" (50 mm) long, with 1/2" (12 mm) of the bolt extending beyond the disc. For suggestions on planning and winding a coil, refer to Chapter 6.

Notice that the ends of the rotor shaft are turned or ground to a 60° point, and that the bearings are center punch marks in the steel brackets.

The rotor may be made of three pieces of sheet metal or laid out and cut from one piece.

The shaft may be made from a long #8-32 (M4) machine screw, or from a piece of round stock. The shaft must be the correct length so it will rotate freely.

Assemble the base and upright and attach the coil.

To accurately position the rotor bearings, check the length of the finished rotor blades and lay out the distance from center of the bearings to the head of the bolt according to your calculations. The clearance between the head of the bolt and the rotor should be 1/32" (1 mm).

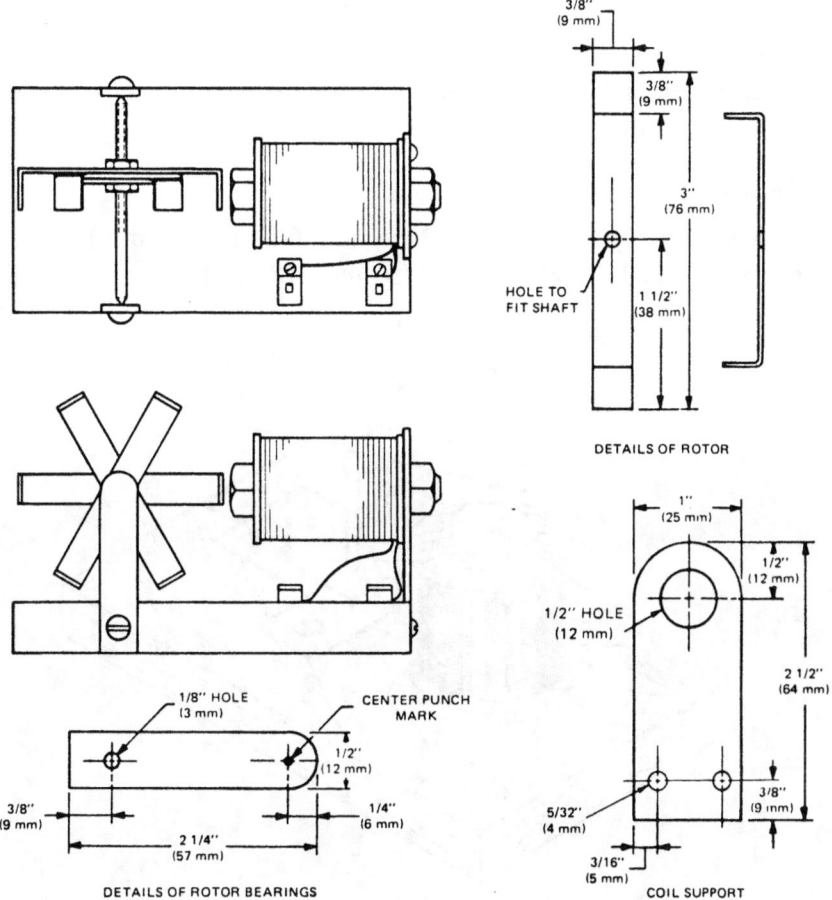

Fig. 16-31

Remember, the coil can be moved toward the rotor poles, but not away from them.

To make an adjustment, the rotor can be raised or lowered.

Assemble the parts, make necessary adjustments, and test.

The rotor poles must be evenly spaced, and the ends should line up with the center of the coil.

Connect the motor to an ac transformer of approximately 6 volts.

Spin the rotor to start the motor running. Several trials may be required as it must rotate at the correct speed before it will continue to run.

You can determine the correct speed by using the following formula:

$$\text{rpm} = \frac{\text{Frequency} \times 120}{\text{Numbers of poles}}$$

Note: The speed of a synchronous motor may be changed by changing the number of poles.

Evaluate.

ALTERNATE DESIGN FOR SYNCHRONOUS MOTOR

The design of this motor, figure 16-32, will permit the use of the solenoid coil described in Chapter 6.

Since the metal parts are easily fabricated, innovations of the basic motor can be readily made.

MATERIALS

1 - Pc. wood, 3/4" x 3 1/2" x 6" (20 mm x 90 mm x 150 mm)
1 - Pc. round metal, 3/16" x 2 3/4" (5 mm x 70 mm)

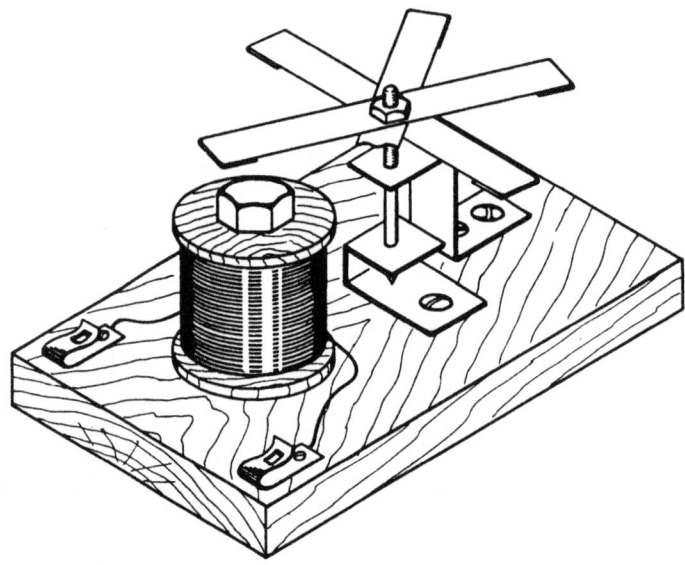

Fig. 16-32

2 - Machine nuts, #10 x 24 (M5)
3 - Pcs. sheet iron, #18, #20 or #22 gauge, 3/8" x 4" (9 mm x 100 mm)
1 - Pc. sheet iron, #18, #20 or #22 gauge, 5/8" x 3 3/4" (16 mm x 95 mm)
1 - Pc. sheet iron, #18, #20 or #22 gauge, 5/8" x 2 1/4" (16 mm x 57 mm)
1 - Machine bolt, 3/8" x 3" (M10 x 75 mm)
1 - Machine bolt nut, 3/8" (M10)
2 - Fahnestock clips
5 - Wood screws, #4 x 1/2" RH
1 - Solenoid coil

PROCEDURE

Study the drawing in figure 16-33 and make the parts according to the specifications on them and in the list of materials.

Notice that the rotor shaft has a point on one end and that it is threaded about an inch on the other end.

After completing all the parts, mount the coil on the base. Place the bearings so the 3/8" (9 mm) fold on the ends of the rotor poles are centered over the bolt head. Adjust the height of the rotor so the poles swing about 1/32" (1 mm) above the bolt head.

1. Check to see that they are all evenly spaced.
2. Attach the coil to a low-voltage ac transformer of about 6 volts.
3. Spin the rotor to start the motor running.

Several trials may be required as it must rotate at the correct speed before it will continue to run.

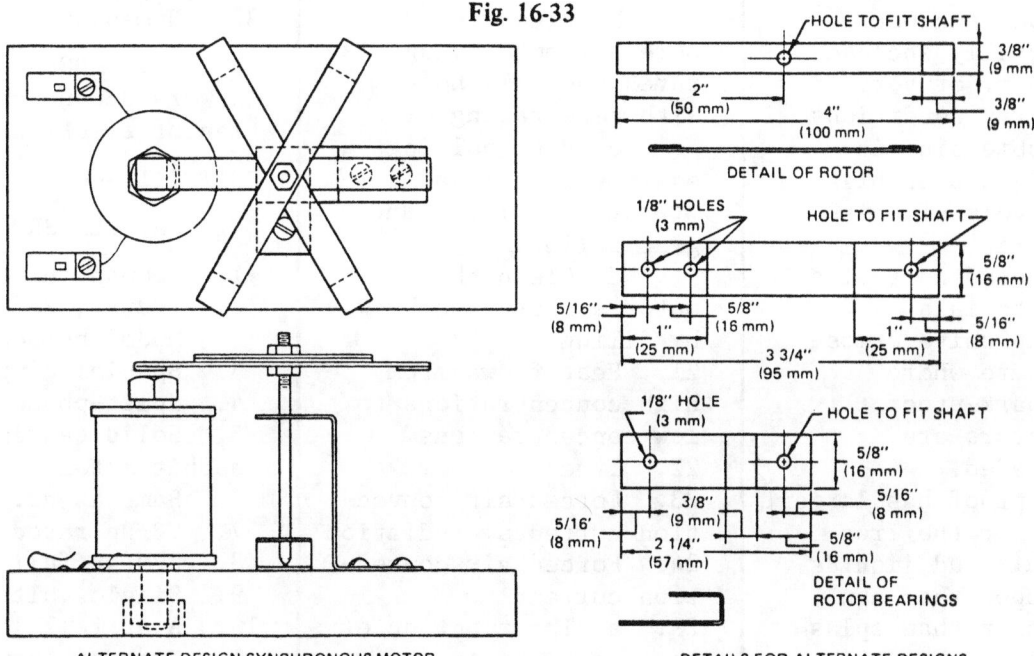

Fig. 16-33

Determine the speed by using the following formula:

$$rpm = \frac{Frequency \times 120}{Number\ of\ poles}$$

Note: The speed of a synchronous motor may be changed by changing the number of poles.

Evaluate.

ANSWERS--CLASSROOM AND LABORATORY EXERCISES

UNIT I

I-A, CLASSROOM

1. Magnetism, electricity
2. D, H, F, B, J, C, G, I, A, E

UNIT II

II-A, CLASSROOM

1. Auxillary
2. Rigid, resilient
3. False
4. Welded mild steel
5. Reduction in vibration
6. One end shield has holes drilled in it to mate with machines having tapped holes for fastening the motor to the machine. An application might be a water pump or fan housing.
7. a. Open enclosure
   b. Totally enclosed
8. a. Dusty or very dirty conditions
   b. Submersion in liquids or high levels of moisture
   c. Around corrosive materials
   d. Explosive atmosphere where spark-proof motors are needed.
9. Drip proof has larger openings for the free flow of air and liquids falling upon them can enter easier than splash proof, which has smaller openings & liquids can not enter as easily by normal splashes.
10. a. (1) b. (2)
    c. (1) d. (2)
    e. (1) f. (3)

II-A (cont'd)

11. Sleeve, anti friction
12. Sleeve
13. Rotor, stator poles
14. Oil film
15. False
16. a. Sealed
    b. Nonsealed whereby lubrication with grease or oil lubricants must be properly and regularly serviced.
17. Sleeve, anti-friction
18. False
19. To physically and mechanically fasten a driver (such as a belt pulley) to the motor's shaft.
20. a. Clean the shaft filing and/or using emery cloth.
    b. Remove set screws from driving fixtures and fill holes with penetrating oil.
    c. Use "pulling equipment" such as a bearing separator and gear puller.
    d. Clean the motor's shaft and keyway by filing.
21. Heat flows from high concentrations to low concentrations.
22. True
23. Forced air convection currents, radiation
24. Forced air convection current
25. a. The function of passing air over the windings.
    b. The ability of the windings to withstand heat.

II-A (cont'd)

26. Varnish insulation
27. Radiation
28. Heat sink
29. a. Breakdown of insulation of the winding conductors.
    b. Overheating the motor due to partial shorts in the windings.
    c. Higher amperage draw and a greater voltage drop.
30. a. Vertical, drill press; b. flat, bench grinder; c. wall, power saw; d. ceiling, any larger industrial piece of equipment solid to the building.
31. True. Unless the sleeve type is designed especially for a mounting position.
32. Nameplate

II-A, LABORATORY
Answers depend on selection of electric motors.

UNIT III

III-A, CLASSROOM

1. Company name
2. Ident., Ref. or Model number
3. Special company number
4. Split-phase
5. Solid bearing out of babbit metal
6. Same as no. 5
7. Y3 Hp rated
8. RPM at full load
9. Single voltage motor
10. A at full load
11. Type of service-1 Ph
12. Cycles per second
13. Service factor
14. Service factor amperes

| III-A, (cont'd) | III-B, (cont'd) | III-B, LABORATORY |

III-A, (cont'd)
15. Locked rotor KVA/HP
16. Efficiency Index
17. Automatic Thermal Overload
18. Connections for low volts
19. Connections for high volts
20. Rating for insulation quality, 130 total temp.
21. Nema standards, fractional Hp
22. Can operate at rated load continuously
23. Amount of overload temperature rise insulation can tolerate
24. Air surrounding motor
25. End bell design
26. Reverse Inst.
27. NEMA number

III-B CLASSROOM
1. 12
2. 16
3. 1 pound feet for both
4. 2 x 96 = 192 oz-ins
5. 192 ÷ 16 ÷ 12 =
   1 x 1750 ÷ 5252
   = .333 Hp
6. .75 x 5252 ÷ 1725 =
   2.28 lb-ft torque
7. 60 x 60 = 3600 ÷ 3
   = 1200 rpm
8. .33 x 1.25
   = .4125 Hp
9. 1200 - 1116 = 84 ÷
   1200 = .07 x 100
   = 7% Slip
10. W = 120 x 5 x 1.0 x 10 = 6000
11. 14.0 x 2 x 1000 ÷ (240 x 1)
    = 116.66 Amp

III-B, (cont'd)
12. (F = C x 9/5 + 32)
    221
    266
    311
    356
13. 145T
14. 32 ÷ 4 = 8"
15. (Unit II) K
    (Fig. 3-6)

III-C, CLASSROOM
A - Table 3-14, Fig. 3-23
1. Code Year
2. No thermal overload
3. Code month
4. Frame, NEMA
5. Cap-start ind-run
6. Speed, 1700
7. Drip proof enclosure
8. Engineering Design
9. Either Electrical or Mechanical modification
10. Serial number of design
11. Factory; Wausau fractional

B - Table 3-15, Fig. 3-23
1. Year
2. Auto Overload
3. Code Month
4. Frame, NEMA
5. Cap-start, cap-run
6. Drip-proof enclosure
7. Rolled steel const.
8a. Engineering design
8b. 6-pole - 1200 rpm
9. modification, electrical
10. modification, mechanical
11. Factory, Wausaw integral

III-A, LABORATORY
Depends on motors selected.

III-B, LABORATORY
1. End bell
2. Starting windings
3. Outside frame
4. Running windings
5. Bearing
6. Squirrel Cage Rotor
7. Motor mount
8. Thrust washer
9. Fan fins
10. Thermal overload
11. Centrifugal switch
12. Nameplate
13. Air vent
14. Assembly bolts
15. Drive end of rotor

III-C, LABORATORY
1. Chuck
2. Bearing
3. Reduction gear
4. Fan
5. Field coil frame
6. Brushes
7. Commutator
8. Switch
9. Handle
10. Case

UNIT IV

IV-A, CLASSROOM
1. AC, DC
2. Generator alternator
3.

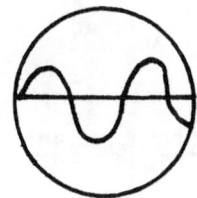

AC

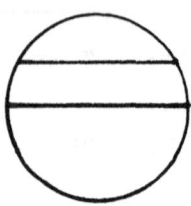

DC
4. 360, +, -, single three
5. speed of rotation strength of magnetic field
6. 120, 50, 30, 15

IV-B, CLASSROOM
1. KVA
2. a. Primary
   b. Secondary
   c. Frame
   d. Left
   e. Right
3. a. Step-down
   b. Step-up
4. 2:1, 1:2
5. 2400   260   12
   120   13   .60
                20:1
6. 2400   260   12
   240   26   1.2
                10:1
7. $\dfrac{EL}{EH} = \dfrac{IH}{IL} = 10$ amp

8. | Volts | Phase |
   |-------|-------|
   | 240 | 1 |
   | 240 | 1 |
   | 240 | 1 |
   | 120 | 1 |
   | 180 | 1 |
   | 240 | 3 |

9. | Volts | Phase |
   |-------|-------|
   | 208 | 1 |
   | 208 | 1 |
   | 208 | 1 |
   | 120 | 1 |
   | 120 | 1 |
   | 120 | 1 |
   | 208 | 3 |

UNIT V------------------
V-A, CLASSROOM
1. a. Electrical service
   b. Physical features
   c. Rotation starting techniques
2. True
3. Stator, rotor
4. Stationary, rotates
5. a. Squirrel cage - Aluminum metal rods supported by end rings run lengthwise & function as electrical conductors. Rotor poles are formed by steel laminations.
   b. Wound - has many strands of copper conductors wound around the shaft & come out to the commutator segments.
6. a. Split-phase
   b. Capacitor
   c. Shaded pole
7. a. Split phase start, induction run
   b. Capacitor start, induction run
   c. Shaded pole start, induction run
   d. Capacitor start, capacitor run
   e. Permanent split capacitor
8. True
9. Permanent magnet lines of flux
10. Current
11. a. Shaping the conductor into coils or loops.
    b. Increasing the number of coils.
    c. Place a metal laminated core inside the coil conductors
    d. Increase the current flow through the coiled conductor

V-A, (Cont'd)
12. False
13. AC current is reversed automatically 60 times per second through generation.
14. Unlike, stator poles, rotor
15. The current flowing through the stator poles establishes a magnetic field and N and S poles. At the same time current is induced into the rotor establishing N and S poles. The N pole of the stator attracts the S pole of the rotor pulling it to align with the pole. At this point, the current in the stator alternates causing the stator poles to change N-S positions thus moving the rotor 180° of rotation. Another change in polarity causes the rotor to turn another 180° getting it back to its original position.
16. Syn Speed =

    $\dfrac{60Hz \times 60 \text{ sec/min}}{\text{pairs of stator poles}}$

    Syn Speed = $\dfrac{3600}{2}$

    Syn Speed = 1800 rpm
17. 90°, 1440°, 900
18. False
19. Slip
20. Starting

V-A, LABORATORY
Answers depend on student selection of machines and motors.

UNIT VI-----------------

VI-A, CLASSROOM

1. Universal
2. Current to stator and commutator
3. Wound rotor
4. Varies with load
5. Large load at low speed or small load at fast speeds.

1. Cap-start Ind.-run
2. Capacitor
3. Squirrel cage
4. 3-6 x rated
5. Hard

1. Split Phase-Start
2. Split Phase
3. Squirrel cage
5. 5-7 x rated
5. Medium

1. Three Phase
2. Windings off-set 120°
3. Squirrel cage
4. Normal to high
5. Medium hard

1. Repulsion-Start, Ind.-run
2. Current induced in rotor
3. Wound rotor
4. 2-4 x rated
5. Very hard

1. Shaded Pole
2. Shading technique
3. Squirrel cage
4. Low
5. Easy

VI-A, LABORATORY
Answers depend on student selection of machines and motors.

UNIT VII-----------------

VII-A, CLASSROOM
(on next page)

VII-A, LABORATORY
Answers depend on student selection of machines and motors.

UNIT VIII---------------------

VIII-A, CLASSROOM
(Ans. VIII and VI)
Split Phase, Medium, 1/6-3/4, 5-7 X rated
Cap Start, Ind Run, Hard. 1/8-10, 3-6 X rated
Perm-Split Cap, Easy, 1/20 to 1, 2-4 X rated
Synchronous, Very Easy Very-Small, Low
Cap. Start. Cap Run, Hard, 2-20, 3-5 X rated
3 Phase, medium-hard, 1/12-200, Normal to high

VIII-A, LABORATORY
Answer depend on motor selected.

UNIT IX---------------------

IX-A, CLASSROOM

a. Efficiency
b. Amps
c. Power Factor
d. Watts
1. The ratio of the actual power used in the circuit in watts, to the apparent power delivered (V x I) Volt-amperes
2. pf = true watts ÷ apparent power (V x I)
3. a. 50-60% or .5-.6
   b. 100% or 1.0

IX-A (cont'd)
3. c. 30-90% or 3-.9
   d. 100% or 1.0
4. No on the above Yes on fractional Hp and some others.
5. Yes
6. 53, 100, (On this motor - but values would not be true on other motors)

IX-B, CLASSROOM

1. a) E d) H g) E
   b) M e) H h) M-
   c) H f) V-   H
              H i) E
3. Only one service as 115
4. Two service as 115.230
5. Shaded pole, universal
7. Permanent-split capacitor
8. Two-value capacitor Permanent-split capacitor
9. Three-phase
10. Soft-start
11. Universal
12. Split-phase
    Shaded pole
    soft-start

IX-A, LABORATORY

1. 10 inch-pound
   160 inch-ounces
   240 in per pound
   20 foot-pound
2. 3.3, 13.2, P
   9.5, 19.0, T
   3.22, 6.44, H
   4.83, 9.66, L
   3.22, 9.66, L

VII-A, CLASSROOM

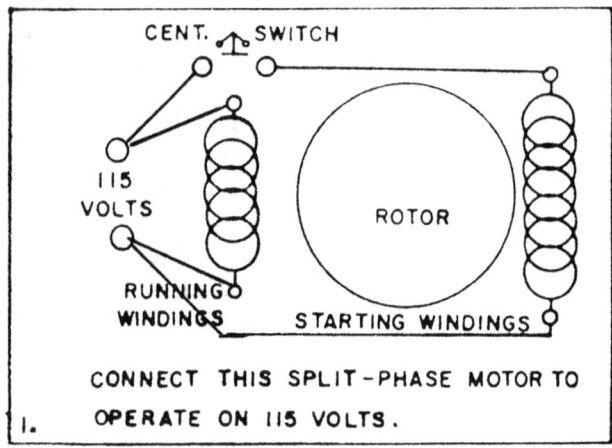

1. CONNECT THIS SPLIT-PHASE MOTOR TO OPERATE ON 115 VOLTS.

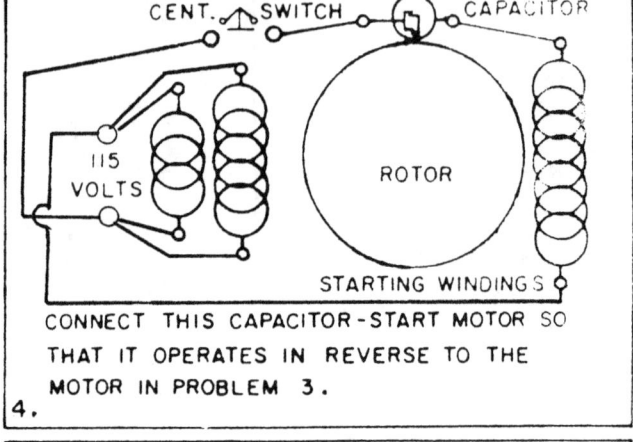

4. CONNECT THIS CAPACITOR-START MOTOR SO THAT IT OPERATES IN REVERSE TO THE MOTOR IN PROBLEM 3.

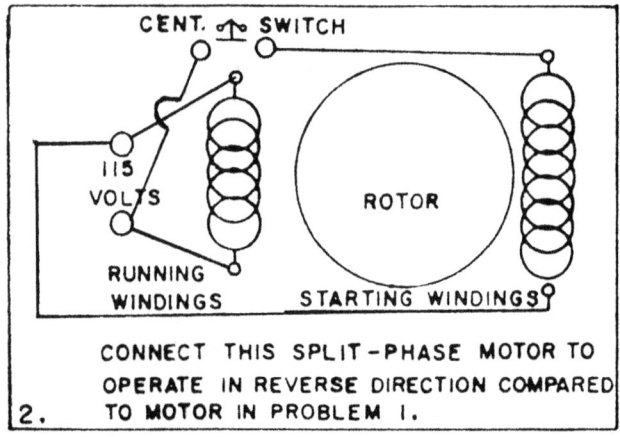

2. CONNECT THIS SPLIT-PHASE MOTOR TO OPERATE IN REVERSE DIRECTION COMPARED TO MOTOR IN PROBLEM 1.

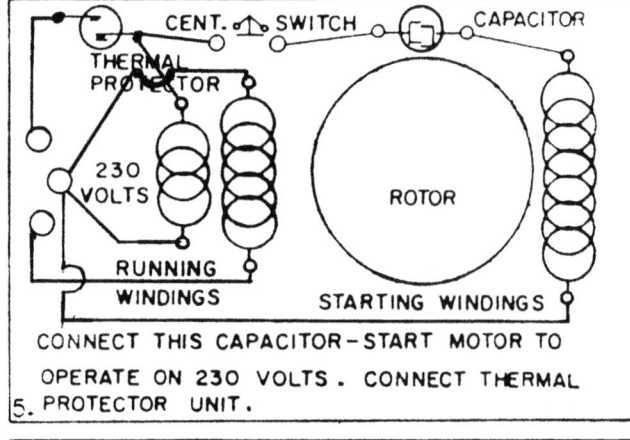

5. CONNECT THIS CAPACITOR-START MOTOR TO OPERATE ON 230 VOLTS. CONNECT THERMAL PROTECTOR UNIT.

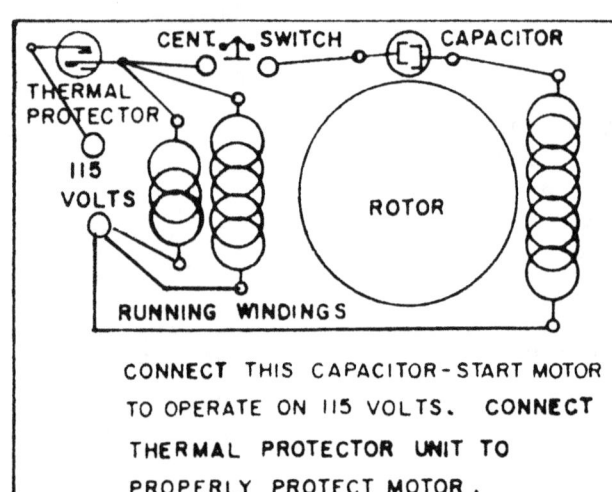

3. CONNECT THIS CAPACITOR-START MOTOR TO OPERATE ON 115 VOLTS. CONNECT THERMAL PROTECTOR UNIT TO PROPERLY PROTECT MOTOR.

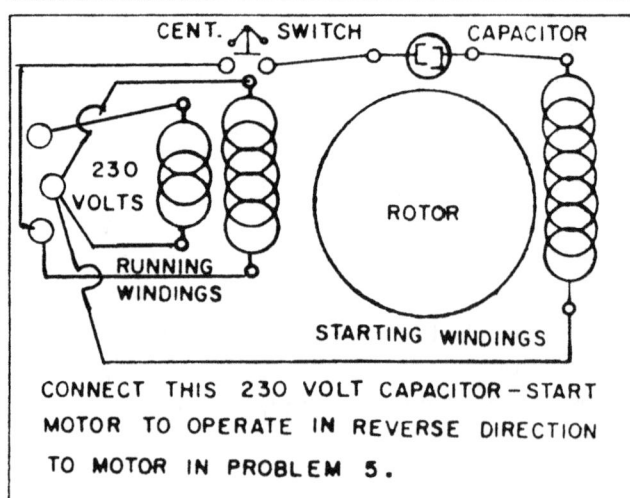

6. CONNECT THIS 230 VOLT CAPACITOR-START MOTOR TO OPERATE IN REVERSE DIRECTION TO MOTOR IN PROBLEM 5.

IX-B, LABORATORY
First table depends on laboratory equipment plus the following answers.
1. Length, load, size
2. 2%, 3%, 4%
3. 2
4. Cost of copper and Labor
5. 12, 8, 8,
   12, 10, 10
   12, 12, 12
   12, 12, 12

If overhead on 125 feet all size 8

IX-C, D, E, LABORATORY
Depends on load used in activity.
   (see sheet).

UNIT X ------------------
1. 12.2
2. 2.8
3. a. 100, 3.6
   b. 100, 2.9
   c. 125, 8.6
      15.1
   d. 10
4. There's a decrease in size, for example, at 45 ampere for 50 feet the size goes from 6- 8- 10.
5. Motor, conductors
6. Table 10.1
   125, 34, 42.0
   125, 5.8, 7.20
7. The fuse will carry a heavy overload while the motor achieves its operating rpm and rated average value.

X-A, (cont'd)
8. Regular
   Time delay or regular
   Time delay
   Time delay
   Regular
   Regular or time delay
   Plug or CB
   Plug or CB
   CB with GFCI
   Cartridge or CB
9. 0.01, 24
10. 10

X-B, CLASSROOM
1. Single pole single throw
2. Single pole double throw
3. Double pole single throw
4. Double pole double throw
5. Triple pole single throw
6. Triple pole double throw
7. N/C, normally closed
8. Normally open

X-C, CLASSROOM
1. It protects the motor and is selected for the motor's full load amperes.
2. P-37
3. 1
4. a. Can be activated by a manual switch
   b. Provides overload protection for the motor
   c. Can be wired to have several switches
5. Normally open
6. Electric current through the coil creates a magnetic field which attracts a metal unit containing one set of contacts which closes the sets of contactors.

X-C, (cont'd)
7. Closed, open, series, parallel
8. N.29
9. a. Can have a switching device at a remote distance from the motor.
   b. Can use a smaller conductor on the control device.
10. a. Thermostat
    b. Humidifier
    c. Limit switch
    d. Photo electric or limit switch
    e. Time clock or repeat cycle timer
    f. Time delay relay

X-D, CLASSROOM
1. N/O Switch
2. N/C Switch
3. N/C Switch
4. N/O Switch
5. Heater coil
6. Motor
7. Battery
8. Resister
9. Photo Electric Coil
10. Diode
11. Capacitor
12. Transformer
13. Rheostat
14. Electromagnet
15. Electric Service Connection

X-E, CLASSROOM
1. Heat, Thermostat
   Moisture, humidstat
   Time, Time clock
    Repeat cycle timer
   Light, Photo Elec.

X-E (cont'd)	XI-A (cont'd)	XI-A, (Cont'd)

1. Pressure, Limit switch Micro/float switch
2. Control, load
3. Control
5. a. Nameplate full-load current
    b. Starter characteristics
    c. Type
6. ⊕, ⊕, parallel, series
 	
X-A, LABORATORY
Completed as student does assignment.
 	
UNIT XI ------------------
 	
XI-A, CLASSROOM
 	
1. Moisture, dust stray oil, air
2. d, b, a, c.
3. An environmentally controlled facility where air is cleaned & kept relatively dust free by filtering. Additionally, workers wear white uniforms.
4. Totally enclosed
5. Positioning of parts whereby they are in one plane or line. In motors, it refers to the general straightness of the drive system.
6. a. Excessive wear on V-belt pulleys.
    b. Damage to thrust washers.
    c. Problems in positioning of starting winding switch
	7. Thrust washers and/or bearings
	8. Anti-friction
	9. Vibration
	10. To transfer power from a motor to a machine
	11. a. 1805.5
	$$\frac{ft.}{min.} = \frac{1\ ft.}{12} \times \frac{12.56}{} \times \frac{1725}{min.} =$$
	1805.5 = 4 x π = 12.56 in
	    b. 3611
	$$\frac{ft.}{min.} = \frac{1\ ft.}{12} \times \frac{12.56}{} \times \frac{3.45}{min.} =$$
	3611 = 4 x π = 12.56 in.
	12. a. High feet per minute linear surface speeds.
	    b. Quiet running
	    c. Inexpensive to replace.
	    d. Good shock absorbing characteristics
	13. 13/32"
	14. Webb multi-v-belt
	15. a. Drill presses
	    b. Woodlathe
	    c. Metal lathe
	16. 4500, yes
	$$\frac{ft.}{min.} = \frac{1\ ft.}{12} \times \frac{9.42}{} \times \frac{1750}{min.} =$$
	1374
	*9.42 ins = pulley circumference of 3 x π
	17. Chain
	18. Worm, spur
	19.
	D x RPM = D x RPM
	(DR) (DR) (DN) (DN)
	6 x 1730 = 2 x RPM
	           (DN)
	RPM = $\frac{6 \times 1730}{2}$ = 5190
	(DN)
		20.
		T x RPM = T x RPM
		(DR) (DR) (DN) (DN)
		15 x 1725 = T x 575
		         (DN)
		T = $\frac{15 \times 1725}{575}$ = 45
		(DN)
		21. Chains, gears belts
		
		XI-A, LABORATORY
		Depends on tools & motor selected.
		
		UNIT XII------------------
		
		XII-A, LABORATORY
		Depends on Laboratory activity.
		
		XII-B, LABORATORY
		Completed as the student does the assignment
		
		XII-C, LABORATORY
		Answers depend on Electric Motor selected by the student

UNIT XIII--------------

XIII-A, CLASSROOM

1. a. Touch b. smell
   c. sight d. hearing
2. a. Loose or disconnected wire; b. blown fuse; c. load too heavy; d. plugged equipment driven by motor.
3. a. disassemble and repair; b. disassemble and repair; c. replace bearings and lubricate d. clean exterior and disassemble to clean inside; e. replace motor or have windings rewound; f. relieve excess to proper level.
4. Arcing
5. a. Universal (AC-DC)
   b. Repulsion start
6. Replace them
7. The motor hums, but will not start and run until he gives the shaft a spin in either direction. Then the motor runs at rated speed.
8. Bad bearings (worn, dirty or lack of lubrication) misalignment of driving system or overload.

NOTES

www.ingramcontent.com/pod-product-compliance
Lightning Source LLC
Chambersburg PA
CBHW060520300426
44112CB00017B/2742